Printed in Mexico

ISBN 978-0-15-362197-0
ISBN 0-15-362197-4

2 3 4 5 6 7 8 9 10   050   16 15 14 13 12 11 10 09 08

Harcourt
SCHOOL PUBLISHERS

Visit *The Learning Site!*
www.harcourtschool.com

What Makes a Home?

Your home is where you live. Around your home is the area where you live. This area includes your home and the plants and animals that live around it. It also includes nonliving things such as rocks, air, and water.

Living things are all around this house. So are nonliving things.

All the living and nonliving things in one place make up an **environment**. Look around the place where you live. How would you describe it? There are many different kinds of environments. Different plants and animals live in each of them.

 MAIN IDEA AND DETAILS What is an environment?

This is another environment with different living and nonliving things in it.

3

Different Environments

Some areas, such as a **desert**, are dry. Many deserts are hot. These environments might have sandy, dry soil and rocks. The animals in these areas are used to living in dry places.

Plants can live in hot dry places too! Have you ever seen a cactus? It stores water in its stem to survive.

Fast Fact

In some deserts, some plants can go without water for years at a time.

These plants survive best in hot, dry areas.

The plants and animals of the rain forest need a wet environment.

A **rain forest** is a wet environment that gets rain almost every day. Not much sunlight reaches the ground through the tall trees. Animals live in the trees or hunt for food on the ground.

 MAIN IDEA AND DETAILS What are two different types of environments?

Environments and Habitats

Most environments have different habitats. A **habitat** is a place where living things have the food, water, and shelter they need to live. One kind of habitat might be a cave. A bat might live in a cave.

These bats use this cave for shelter. They eat insects that live nearby.

A pond is a good habitat for many kinds of animals.

A **pond** is a freshwater habitat. Fish live in the water of the pond. They eat plants that grow in the pond, and they hide behind underwater rocks. A duck at the pond makes a nest in plants near the shore. It eats fish and plants from the pond.

 MAIN IDEA AND DETAILS What is a habitat?

Adapting to Survive

Over time, animals and plants change. They **adapt**, or change, so they can live in the area around them.

Most living things have adapted to live in one environment. A desert lizard can get rid of body heat quickly. It can not survive in a cold mountain forest.

Can you see this lizard? Over time, lizards have adapted so they can hide from animals that might eat them.

A grizzly bear is adapted to cold climates. It grows fur, stores fat, and sleeps through winter. The desert is too hot for a bear to live in. It would not be able to find nuts or fish to eat. A bear would also need more water to survive.

Fast Fact

Some birds are adapted to water. They have webbed feet to help them swim. Oily feathers keep them dry.

 MAIN IDEA AND DETAILS Why can most animals not survive in different environments?

Bears are adapted to living in cold climates. They grow thick fur and layers of fat.

Changing Environments

Sometimes people change the environment. They drain lakes or cut down forests to build homes. These changes harm the habitats.

Other changes help the environment. People replant trees where other trees were cut down for wood. People set up places called refuges where plants and animals can be safe.

 MAIN IDEA AND DETAILS What are the different ways people change the environment?

Many more plants and animals used to live in the Everglades.

Parts of the Everglades have been drained and many habitats destroyed.

Summary

An environment is made up of all the living and nonliving things in a place. There are many different environments. Each environment has habitats in it. Over time, animals and plants have adapted to live in habitats. Some people change the environment to help plants and animals.

Glossary

adapt To change. Animals and plants adapt over time to live in their environment. (8, 9, 11)

desert A dry environment that gets little rain (4, 5, 8, 9)

environment All the living and nonliving things in a place (3, 4, 5, 6, 7, 8, 10, 11)

habitat A place where living things have the food, water, and shelter they need to live (6, 7, 11)

pond A small freshwater environment (6, 7)

rain forest A wet environment that gets rain almost every day (5)